P9-APG-126

U.S. SPECIAL OPS FORCES

INSIDE THE NAVY SEALS

HOWARD PHILLIPS

PowerKiDS press

NEW YORK

Published in 2022 by The Rosen Publishing Group, Inc.
29 East 21st Street, New York, NY 10010

First Edition

Editor: Greg Roza
Designer: Rachel Rising

Portions of this work were originally authored by Drew Nelson and published as *Navy SEALs*. All new material in this edition was authored by Howard Phillips.

Photo Credits: Cover MILpictures by Tom Weber/The Image Bank/Getty Images; cover, pp. 1–32 CRVL/Shutterstock.com; cover, p. 1 Zsschreiner/Shutterstock.com; cover, pp. 1, 3, 21, 30, 31, 32 Massimo Saivezzo/Shutterstock.com; p. 4 pamelasphotopoetry/Shutterstock.com; pp. 5, 11 Getty Images/Handout/Getty Images News/Getty Images; pp. 7, 9, 11, 15, 17, 21, 27 Khvost/Shutterstock.com; p. 7 U.S. Navy/Handout/Archive Photos/Getty Images; p. 9 PhotoQuest/Contributor/Archive Photos/Getty Images; pp. 13, 22, 28 Stocktrek Images/Getty Images; p. 15 Getty Images/Staff/Getty Images News/Getty Images; p. 17 Wathiq Khuzaie/Stringer/Getty Images News/Getty Images; p. 18 https://commons.wikimedia.org/wiki/File:SEAL-TEAM4.jpg; p. 18 https://commons.wikimedia.org/wiki/File:SEAL-TEAM5.jpg; p. 18 https://en.wikipedia.org/wiki/File:Logo_Naval_Special_Warfare_Development_Group.svg; pp. 19, 21, 23 Richard Schoenberg/Contributor/Corbis News/Getty Images; p. 24 Handout/Getty Images News/Getty Images; p. 25 U.S. Navy/Handout/Getty Images News/Getty Images; p. 27 https://commons.wikimedia.org/wiki/File:Monsoor.jpg; p. 29 Jim Sugar/The Image Bank Unreleased/Getty Images.

Some of the images in this book illustrate individuals who are models. The depictions do not imply actual situations or events.

Library of Congress Cataloging-in-Publication Data

Names: Phillips, Howard, 1971- author.
Title: Inside the Navy Seals / Howard Phillips.
Description: New York : PowerKids Press, [2022] | Series: U.S. Special Ops forces | Includes index.
Identifiers: LCCN 2020052580 | ISBN 9781725328990 (library binding) | ISBN 9781725328976 (paperback) | ISBN 9781725328983 (6 pack)
Subjects: LCSH: United States. Navy. SEALs–History–Juvenile literature. | United States. Navy–Commando troops–Juvenile literature.
Classification: LCC VG87 .P54 2022 | DDC 359.9/84-dc23
LC record available at https://lccn.loc.gov/2020052580

Manufactured in the United States of America

CPSIA Compliance Information: Batch #BSPK22. For further information contact Rosen Publishing, New York, New York at 1-800-237-9932.

CONTENTS

MEET THE NAVY SEALS

The U.S. Navy SEALs are **elite** sailors trained to carry out special missions and **covert** operations all over the world. Their name comes from all the different kinds of places they can work. "SEAL" stands for "SEa, Air, and Land." SEALs are part of a division of the navy called the Naval Special Warfare community.

Navy SEALs are deployed around the world in small, well-trained teams to carry out important tasks. Many times, they work quietly and at night to carry out their missions in secret. The SEALs work to keep the United States safe.

ALL SEALS WEAR THE SPECIAL WARFARE INSIGNIA, ALSO CALLED THE SEAL TRIDENT. A SOLDIER WEARING THE TRIDENT HAS PASSED BASIC UNDERWATER **DEMOLITION**/ SEAL SCHOOL TO BECOME A NAVY SEAL.

BEFORE THE SEALS

The SEALs weren't the first special land and sea forces for the U.S. Navy. In 1942, during World War II, the army and navy started jointly training a beach **reconnaissance** force called the Scouts and Raiders. Their job was to secretly go to a beach before the rest of the soldiers got there. The Scouts and Raiders figured out enemy positions and planned attacks.

In 1943, the navy created the Underwater Demolition Teams (UDTs). These special groups wore swimsuits, fins, and facemasks, and were trained to plant bombs on enemy targets underwater. UDTs also fought during the Korean War and the Vietnam War.

OTHER EARLY NAVY SPECIAL FORCES

DURING WORLD WAR II, NAVAL COMBAT DEMOLITION UNITS WERE SAILORS TRAINED IN BOTH BEACH AND ABOVE-WATER EXPLOSIVES. OFFICE OF STRATEGIC SERVICES OPERATIONAL SWIMMERS WERE TRAINED TO SWIM INTO AND OUT OF ENEMY WATERS ON RECONNAISSANCE AND COMBAT MISSIONS.

UDT SOLDIERS WERE NICKNAMED "FROGMEN" BECAUSE THEY WERE ADEPT AT UNDERWATER MISSIONS. THIS TERM IS STILL USED FOR NAVY SEALS TODAY.

JFK AND THE SEALS

President John F. Kennedy wanted a Special Forces group that could complete unusual missions. In 1961, he asked the military to form new teams that could carry out secret operations in or near rivers and oceans. In January 1962, the U.S. Navy formed SEAL Teams One and Two. These two teams were made up of members of Underwater Demolition Teams.

The first war the Navy SEALs fought in was the Vietnam War. At first, they just gave advice to other members of the military. The Navy SEALs began their first active mission in February 1966.

THE TOP FROG!

THE ACTIVE-DUTY SEAL WHO'S SERVED THE LONGEST IS GIVEN THE TITLE OF "BULL FROG." HE ALSO GETS A TROPHY TO DISPLAY FOR THE ENTIRE TIME HE'S THE BULL FROG

THIS PHOTOGRAPH FROM NOVEMBER 1967 SHOWS NAVY SEAL TEAM ONE IN AN ASSAULT BOAT ON THE BASSAC RIVER SOUTH OF SAIGON, VIETNAM.

SEAL OPERATIONS

After the Vietnam War, many SEALs continued to work in Vietnam. They gave advice to the military there until 1973. Then, in May 1983, all members of the Underwater Demolition Teams became Navy SEALs or part of the Swimmer Delivery Vehicle Teams (SDVTs). The SDVTs, now called SEAL Delivery Vehicle Teams, get the SEALs where they need to be.

Between 1983 and 2012, the SEALs carried out missions all over the world. They took part in Operation Earnest Will in the Persian Gulf and Operation Desert Storm in the Persian Gulf. In 2011, Navy SEALs Team Six located and **assassinated** al-Qaeda leader Osama bin Laden.

ON MAY 2, 2011, NAVY SEAL TEAM SIX LANDED TWO HELICOPTERS NEAR THE COMPOUND OF OSAMA BIN LADEN IN PAKISTAN. USING NIGHT-VISION GOGGLES, THEY SNUCK INTO THE COMPOUND IN THE MIDDLE OF THE NIGHT. THE SEALS FOUND BIN LADEN ON THE THIRD FLOOR. AFTER A SHORT FIGHT, THE SEALS REPORTED THAT THEY HAD KILLED THE TERRORIST LEADER OF AL-QAEDA.

IN THIS PHOTOGRAPH FROM 2003—DURING THE U.S. MISSION AGAINST **TERRORISM** KNOWN AS OPERATION ENDURING FREEDOM—SEALS TRAIN IN **RAPPELLING** FROM A HELICOPTER TO THE DECK OF AN AIRCRAFT CARRIER.

SEAL MISSIONS

Navy SEAL missions are sorted into four main categories: direct action, special reconnaissance, foreign internal defense, and counterterrorism. In direct action, SEALs attack enemy targets. In special reconnaissance missions, SEALs sneak into enemy territory and find out information, such as where the enemies are, what they're doing, and what kind of forces they have.

In foreign internal defense, the SEALs train and help soldiers in other countries act when their enemies attack. In counterterrorism, the SEALs follow the activities of terrorist groups and act directly against them to prevent terrorist attacks.

THE SEALS MUST BE PREPARED FOR ANY TYPE OF MISSION AT SEA, IN THE AIR, OR ON LAND. SHOWN HERE, A NAVY SEAL TEAM LAUNCHES A SEAL DELIVERY VEHICLE (SDV) FROM THE BACK OF A SUBMARINE. THIS SUBMERSIBLE IS USED FOR DIRECT ACTION AND UNDERWATER RECONNAISSANCE MISSIONS.

OPERATION ENDURING FREEDOM

After September 11, 2001, when the United States suffered three terrorist attacks, U.S. Navy SEALs were sent into Afghanistan. In fact, the first high-ranking officer in Afghanistan after military operations began was a Navy SEAL.

The mission in Afghanistan was called Operation Enduring Freedom. The SEALs and other U.S. special forces completed more than 75 special reconnaissance and direct action missions. This included searching for terrorists trying to leave the country and getting rid of more than 500,000 pounds (227,000 kg) of weapons and explosives. Operation Enduring Freedom came to an end in 2014, although some U.S. forces remained in Afghanistan to help local forces rebuild.

MEDALS OF HONOR

THE MEDAL OF HONOR IS THE HIGHEST AWARD A U.S. SERVICE MEMBER CAN RECEIVE. IT'S GIVEN TO THOSE WHO HAVE SHOWN ACTS OF **VALOR** IN COMBAT. THE U.S. NAVY HAD THREE MEDAL OF HONOR RECIPIENTS DURING OPERATION ENDURING FREEDOM, AND THEY WERE ALL SEALS: LIEUTENANT MICHAEL P. MURPHY, SENIOR CHIEF SPECIAL WARFARE OPERATOR EDWARD C. BYERS JR, AND MASTER CHIEF PETTY OFFICER BRITT SLABINSKI.

NAVY SEALS CARRIED OUT DIRECT ACTION AND COUNTERTERRORISM MISSIONS DURING THE WAR ON TERROR IN THE EARLY 2000S. THIS PHOTOGRAPH SHOWS ONE OF 70 CAVES IN AFGHANISTAN, USED AS BASES BY AL-QAEDA FORCES, THAT SEALS DESTROYED.

OPERATION IRAQI FREEDOM

In 2003, Operation Iraqi Freedom was a U.S.-led mission to overthrow the Iraqi government and its leader Saddam Hussein. This mission used more SEALs than any other in the history of the Naval Special Warfare group.

The SEALs carried out many military actions during their time in Iraq. These included securing places that produced oil, clearing waterways to allow the transport of aid to the people of Iraq, attacking possible terrorist locations, and capturing most-wanted enemies. On April 1, 2003, a team of Navy SEALs and Army Rangers rescued army Private First Class Jessica Lynch from enemy forces in Nasiriyah, Iraq.

JESSICA LYNCH WAS AN ARMY UNIT SUPPLY SPECIALIST DURING THE WAR IN IRAQ. HER VEHICLE WAS ATTACKED BY ENEMY FORCES, AND LYNCH WAS SERIOUSLY INJURED. SHE AND OTHER U.S. SOLDIERS WERE CAPTURED AND BECAME POWS, OR PRISONERS OF WAR. LYNCH BECAME THE FIRST AMERICAN POW TO BE RESCUED SINCE WORLD WAR II.

ON APRIL 9, 2003, U.S. SOLDIERS HELPED IRAQI FORCES TO PULL DOWN A STATUE OF SADDAM HUSSEIN AT AL-FARDOUS SQUARE IN BAGHDAD.

DUTY STATIONS

SEAL teams are stationed at three duty stations across the country. The station in Coronado, California, is home to the Naval Special Warfare Center and Naval Special Warfare Command. The group that runs SEAL training is also there, as well as SEAL Teams One, Three, Five, and Seven. SEAL Team Seventeen, a reserve unit composed of sailors not on active duty, is stationed there too.

Little Creek, Virginia, is home to SEAL Teams Two, Four, Eight, and Ten, as well as the reserve unit SEAL Team Eighteen and SEAL Delivery Team Two. SEAL Team Six is located in nearby Virginia Beach. The third duty station is in Pearl Harbor, Hawaii. This is where SEAL Delivery Vehicle Team One is located.

SOLDIERS TRAIN TO BECOME NAVY SEALS IN CORONADO, CALIFORNIA, DURING A VERY TOUGH WEEK OF TRAINING KNOWN AS "HELL WEEK."

TOUGH TRAINING

The training to become a SEAL lasts for 2½ years. It's considered the hardest military training anywhere in the world.

First, trainees must go to SEAL Prep School and take a difficult fitness test. In the school, they train to improve their test performance for an even harder fitness test they must complete in order to continue with SEAL training. If they pass this second test, trainees enter BUD/S school, a 24-week program that teaches **stamina** and leadership skills. There, they have three different training sessions. Each takes seven weeks and includes physical conditioning, combat diving, and land warfare.

SEALS ARE TRAINED TO WORK WELL AS A TEAM.

MAKING THE CUT

SEAL TRAINING IS SO TOUGH MANY CANDIDATES DON'T MAKE THE CUT. IN ORDER TO EVEN BEGIN THE TRAINING, CANDIDATES MUST COMPLETE THE FOLLOWING TASKS:

- SWIM 500 YARDS (457.2 M) IN 12 MINUTES, 30 SECONDS
- DO 42 PUSH-UPS IN 2 MINUTES
- DO 50 SIT-UPS IN 2 MINUTES
- DO 6 PULL-UPS WITH NO TIME LIMIT
- RUN 1.5 MILES (2.4 KM) IN 11 MINUTES

Candidates who successfully complete BUD/S school go to a three-week session of Parachute Jump School, where they're trained to jump from planes. Once this is finished, they go into SEAL Qualification Training, a 26-week program at the duty station at Coronado. There, trainees learn skills specific to the SEALs and their missions, including how to endure cold water, sea operations, combat swimming, and how to fight in small spaces.

After trainees have graduated from SEAL Qualification Training, they spend up to 18 months taking part in advanced training for different positions, such as medical officer, communications expert, and explosives expert.

THESE SEAL CANDIDATES HAVE PASSED BUD/S, AND WILL SOON GO THROUGH PARACHUTE JUMP SCHOOL.

ON THE TEAM

About 1,000 men begin SEAL training at the BUD/S school every year. However, only between 200 and 250 complete the training and become SEALs. People who enter the school but can't complete it on the first try are given a different job with the navy. They're allowed to apply again after two years.

There are only about 2,500 active-duty SEALs at any time, and about 500 of them are officers. The eight SEAL teams are divided into platoons. A platoon usually has 16 members, but could be broken into smaller units called squads and elements. In addition, there are two SEAL Delivery Vehicle Teams.

SOME SEALS BECOME **SNIPERS**, SOME BECOME LANGUAGE EXPERTS, WHILE OTHERS BECOME JUMPMASTERS, WHICH IS THE PERSON IN CHARGE OF A TEAM OF SOLDIERS PARACHUTING FROM A PLANE.

FAMOUS NAVY SEALS

Many SEALs have gone on to become famous after leaving the military. Former SEAL Jesse Ventura, who fought in the Vietnam War, became a professional wrestler known as "The Body." Ventura went on to become an actor, radio show host, and the governor of Minnesota!

Rudy Boesch was on the first season of the TV show *Survivor* and finished in third place. Former SEALs Harry Humphries and Chuck Pfarrer work in the movie business, giving advice and writing. Two former SEALs, William M. Shepherd and Chris Cassidy, even became astronauts and went into space! Both Dick Couch and Richard Marcinko went on to become famous authors.

AMERICAN SNIPER

FROM 1999 TO 2009, CHRIS KYLE SERVED FOUR TOURS OF DUTY IN IRAQ AS A SEAL SNIPER. HE IS KNOWN AS THE MOST **LETHAL** SNIPER IN AMERICAN HISTORY. KYLE WROTE HIS AUTOBIOGRAPHY—*AMERICAN SNIPER*—IN 2012 AND IT BECAME A HUGE SUCCESS. THE BOOK BECAME A POPULAR MOVIE IN 2014, STARRING BRADLEY COOPER AS CHRIS KYLE.

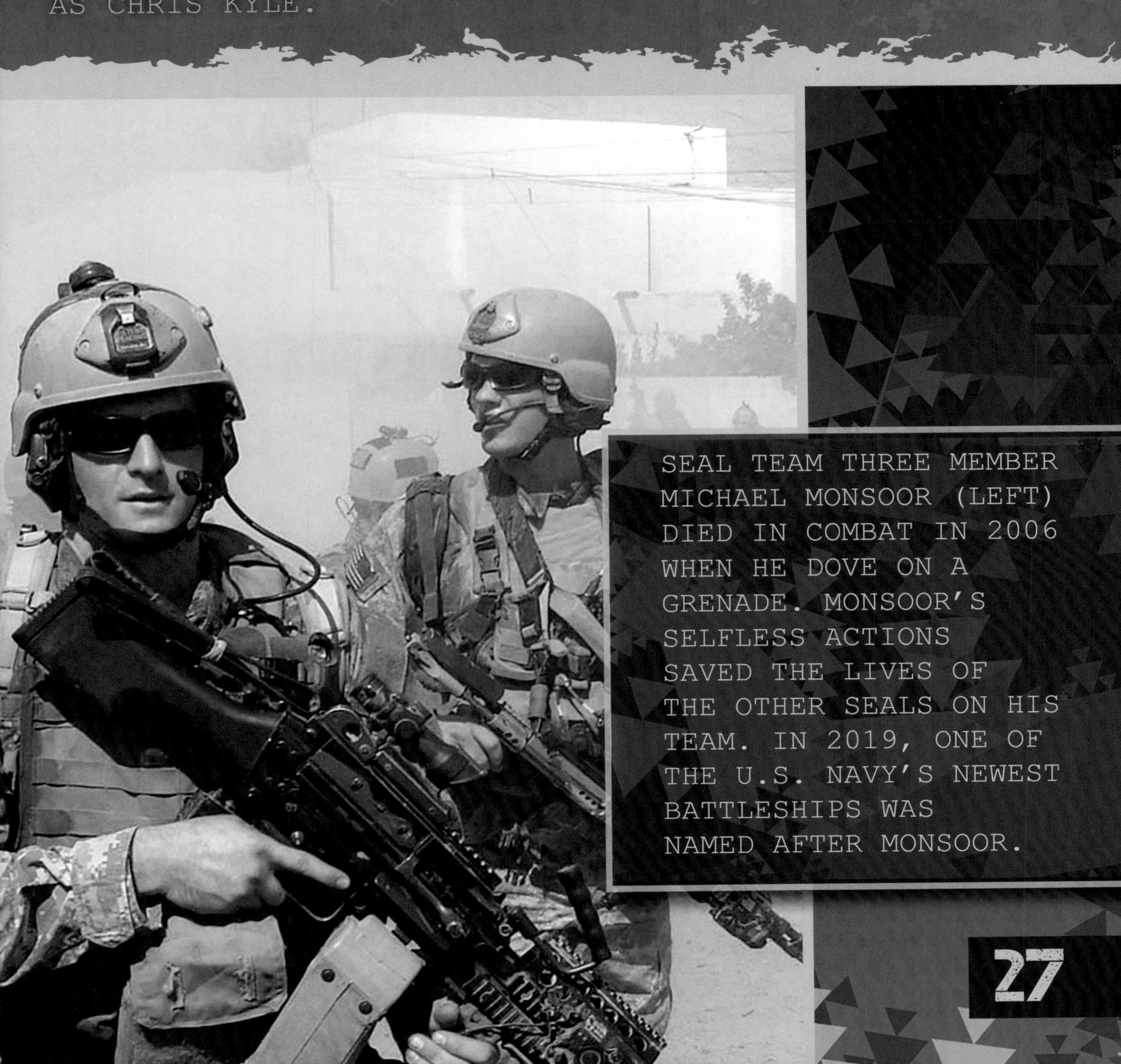

SEAL TEAM THREE MEMBER MICHAEL MONSOOR (LEFT) DIED IN COMBAT IN 2006 WHEN HE DOVE ON A GRENADE. MONSOOR'S SELFLESS ACTIONS SAVED THE LIVES OF THE OTHER SEALS ON HIS TEAM. IN 2019, ONE OF THE U.S. NAVY'S NEWEST BATTLESHIPS WAS NAMED AFTER MONSOOR.

NEVER QUIT

In early May 2011, SEAL Team Six completed Operation Neptune Spear, the secret mission to capture or kill Osama bin Laden. The U.S. Navy SEALs will continue to carry out special missions at sea, in the air, and on land for years to come. They operate secretly all over the world to make sure all Americans are safe.

There is a special code that all SEALs follow in life. This is called the SEAL **Ethos**. In this code, SEALs promise to be loyal, never quit, serve with honor, and stop America's enemies. This ethos guides SEALs in their work, and in their personal lives.

SEALS ARE PREPARED FOR COMBAT IN ANY PLACE AND AT ANY TIME.

GLOSSARY

assassinate: To kill a usually important person by surprise attack.

covert: Made or done secretly.

demolition: The act of working with explosives to destroy things.

elite: Superior in talent and ability.

ethos: The guiding beliefs of a person, group, or organization.

lethal: Deadly.

rappel: To slide down a rope from a high location.

reconnaissance: Military activity in which soldiers are sent to find out information about an enemy.

sniper: Someone who is good at shooting targets from a hiding place.

stamina: The ability or strength to keep doing something for a long time.

terrorism: The use of violence to scare people as a way of achieving a political goal.

valor: Courage or bravery.

FOR MORE INFORMATION

BOOKS

Boothroyd, Jennifer. *Inside the U.S. Navy*. Minneapolis, MN: Learner Publications, 2017.

Morey, Allan. *U.S. Navy*. Minneapolis, MN: Jump! Inc., 2020.

WEBSITES

How the Navy SEALs Work

science.howstuffworks.com/navy-seal.htm

Check out this thorough website to learn more about the Navy SEALs.

SEAL Ethos

www.nsw.navy.mil/NSW/SEAL-Ethos/

Visit this website to read the official Navy SEAL Ethos.

Publisher's note to educators and parents: Our editors have carefully reviewed these websites to ensure that they are suitable for students. Many websites change frequently, however, and we cannot guarantee that a site's future contents will continue to meet our high standards of quality and educational value. Be advised that students should be closely supervised whenever they access the internet.

INDEX